The Biotech Blueprint

From Aspiration to Achievement

By: Te'Jon McRae

Table of Contents

Foreword 1

A Bird's Eye View 4

From Concept to Creation 12

Baby Steps into Biotech 21

Forging A Career 27

Develop Your Learning Style 38

Getting a Good Start 42

Crafting A Niche 47

Seeing Is Believing 56

Everyone Needs Support 62

Persevere… and Then Persevere Some More 68

Broaden Your Knowledge 74

Parting Thoughts 80

Works Cited 84

FOREWORD

Before I begin, I first want to give praise and glory to God! Through Him, all things are possible and no challenge is too great. May His light radiate through my words and impart you with a knowledge or wisdom that you didn't have prior to reading this book.

I want to give thanks to my Mom (Moniquetta), Dad (Rodney), Grandma(Deloris) and Grandpa (Doug). May you rest in peace and may your sacrifices, love, and lessons give me the capacity to guide and support a new generation. Every moment I spent with you was cherished and I love you always.

My next outpouring of love goes to my family: my sisters Chenae and Kyla and my cousins Ciara, DJ and Paris. We all grew up together and I still look back so fondly on the days we would get in trouble together. Finding ways to get out of trouble or taking the fall together still bring a smile to my face. I love that we've remained with each other this entire time and can't wait to see how we'll build our futures, as well as the futures of Jaiden, Jaida, Sa'Maya, Iyana, Sire and Samir.

To the Smith family, I want to graciously give thanks to the love and support you all have presented me with over the years. Shenetta, Gregory, Layne, Jared and Khayla- each of you have touched my life and heart in ways that I cherish deeply and profoundly. We went from strangers to family, and I can't imagine where my life would be without having met you guys. May our bonds only grow stronger and become more enriched.

To Dalvin and Ashley, the love, structure and fun we had while we were in the Bay is a core memory for me. Whether watching a movie, concocting business strategies, exploring the Bay or just imparting wisdom, I was honored to have been welcomed into your homes and your hearts. And to be able to be there for Hudson and Sora and be their uncle Tay makes my heart skip a beat.

When I first moved to the Bay, I wanted to find a way to give back to my community and others. I've always felt strongly about making the world a slightly better place. I was forged by the pain and trauma of such heavy loss while I was a kid. This pain made me determined to be a powerful force for others, striving to become a pillar that others could lean on- a resource.

What I set out to do was to impart my knowledge and insights from having worked in the Biotech industry. Whether from my personal experiences, research, or tribal knowledge, I saw a dire need to address gaps in the industry for both new and aspiring individuals to better develop their paths for growth. I pray that my words carry some worth for you and that I can, in some meaningful way, contribute to your future endeavors.

Life:

Gracious spring,

bringing forth warmth and beauty The arbiter of life

Courageous Summer

children laughing, food simmering, sunlight beaming

Breezy autumn,

full of fallen leaves and pumpkin spice Breathing new life into a weary world

Dreary winter,

with sharp winds and heavy snow Families huddling together for love and warmth

Gracious spring, the cycle begins anew

- J. D. Rae

A Bird's Eye View

"You never really understand a person until you consider things from his point of view... Until you climb inside of his skin and walk around in it."

— Harper Lee, To Kill A Mockingbird

A fantastic place to begin is offering insight into biotechnology. What is it? What role does it play in the world? What type of individuals typically join it? When I first got into the biotech industry, I had absolutely no idea what it was. My vision was focused on finding a well-paying job out of college that utilized my Biochemistry degree. It wasn't until I began to understand the significance of what I was doing that full weight of my work dawned on me.

Biotechnology is a sector that combines some of the best parts of science -biology, chemistry and technological achievements- to conquer different problems that plague the world. Some problems are very obvious and call for immediate actions. Cancer is a very identifiable problem that still plagues the world, but even things like crop failure or vaccines are applicable. Other common applications of biotech can include pharmaceuticals, agriculture, medical devices and even beer making.

Biotech, Pharma ... What's the difference?

You'll often hear the terms Biotech and Pharmaceutical thrown around, but what do they mean? How are they different? Are they different? The truth is a bit muddier than you might imagine.

We can start with Biotech first. In an overall sense, a biotech company is any company that develops a product from biological organisms. Biotechnology has actually been a part of human civilization since the first beer products made use of yeast for its alcohol production. Another ancient application of biotech is the experimentation and creation of more resilient and productive crops. More recent applications see the development of plastics or Cell Therapy, a medical application to develop treatment drugs from a person's immune system. Cell Therapy in particular is a huge subset due to its innovative approaches to health problems that are becoming ever more prevalent in a world full of pollution, microplastics and unhealthy food options. Some examples of a Biotech company includes BioMarin and Roche.

Next, we can touch on Pharmaceuticals or pharma for short. Pharma also focuses on the creation of medications and products, but instead of being from biological organisms, they come from synthetic compounds and chemicals. Many of these products will have a chemical structure you can find online, such as Acetaminophen and Ibuprofen. Some big-name Pharma companies include Roche, Johnson & Johnson and Eli Lilly.

If you noticed, I put Roche twice and the reason for that is simple: there is a ton of overlap between Pharma and Biotech. Many companies that have their hands in the Pharma space will also have them in Biotech and vice versa.

They're able to diversify their revenue streams and investments this way. So, while there are distinct differences between the terms, in practice there's a large overlap between the companies that they apply to. Even with all of this knowledge, it can be hard to understand how exactly something can go from a concept to a fully realized product. In the next chapter, we will focus on illuminating the general Biotech pathway to commercialization. Below, we look at the strength and statistics of the industry.

A glimpse into the industry

For a majority of this book, I talk about the Bio-Pharmacy space as that's what I'm most exposed to. As you can see in **Figure 1** almost half of the Biotech market space is Bio-Pharma related followed by the others (*Precedence Research* 2024). There's been a rising prevalence of illnesses and sicknesses in the world. Diseases have become more common and resistant to drugs. As pointed out by Rachel Baker, "Selection for drug resistance now occurs worldwide, and antibiotic resistance has and will evolve repeatedly" (*Baker et al.* 2021).

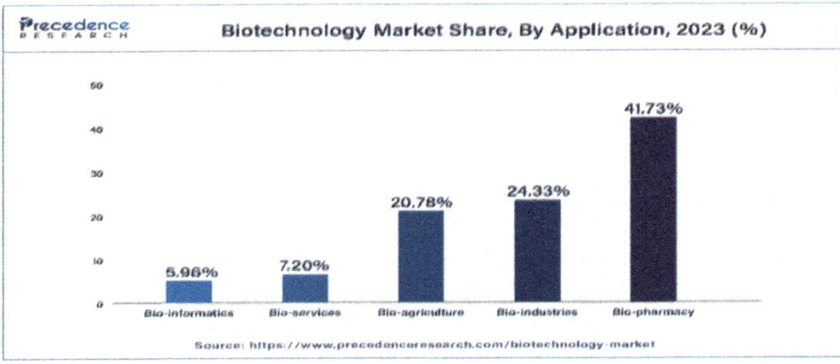

Figure 1: Biotech Industry Distribution

Looking towards further financial tides, the US Biotech market is projected to grow over the next 10 years as seen in **Figure 2** (*Precedence Research* 2024). This growth reflects only the Biotech market, not the Pharmaceutical market which has a market size closer to $600 billion USD. The projected growth of Biotech will see increased demand for products, more revenue opportunities, and more jobs generated. Coupled with the rise in disease, hardships caused by Climate Change, and pollution, the need to have innovative solutions to a wide range of problems will only become more apparaent.

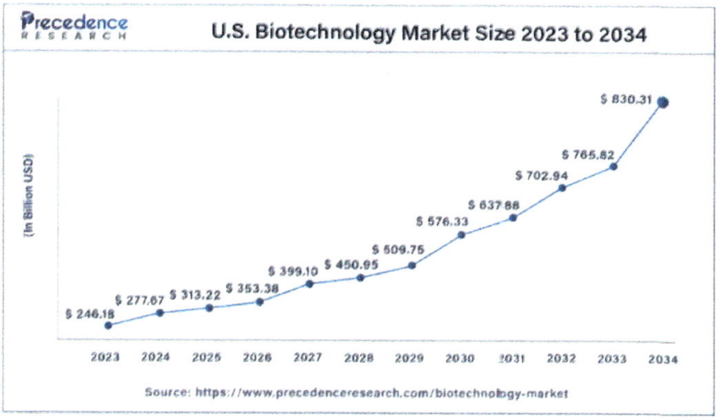

Figure 2: U.S. Biotech Market Share

Something that stood out to me strongly in my research was the demographic make up of the Biotech industry. As a Black man in the industry, I could anecdotally see it from my day-to-day interactions and workspaces. However, **Figure 3** below really highlights the disparities present in the industry. A 2022 report conducted by Bio.org- the world's largest advocacy group for biotech companies and academic institutions- showed that only 6% of the Biotech workforce identified as Black (*Bio.org* 2022). While the sample size wasn't large, the numbers aren't too far off from other reports. I say this not to discourage any minority groups from working in Biotech, but rather to encourage you. A healthy and thriving workplace depends on people of different backgrounds (ethnic, racial, national, education or otherwise). Our differences make us stronger!

Race and ethnicity	Total employees (n=28)	Exec (n=31)
White	56%	72%
Asian	21%	15%
Hispanic/Latinx	7%	4%
Black	6%	3%
2+ races	3%	0.3%
Native Hawaiian/Pacific Islander	0.5%	0%
Native American/Alaskan Native	0.1%	2%
Not disclosed	6%	4%

Figure 3: Biotech Demographics

A discussion about the Biotech industry also requires a discussion about the pay within it. In **Figure 4**, a study conducted by Lightcast Analyst compiled data from across the industry and detailed the average annual salaries of workers. Different positions garnered different salaries, with some going well into the 6-figure range (Joubert 2024). The Biotech industry is not only fast growing, but its pay is keeping pace for the life-saving work being done.

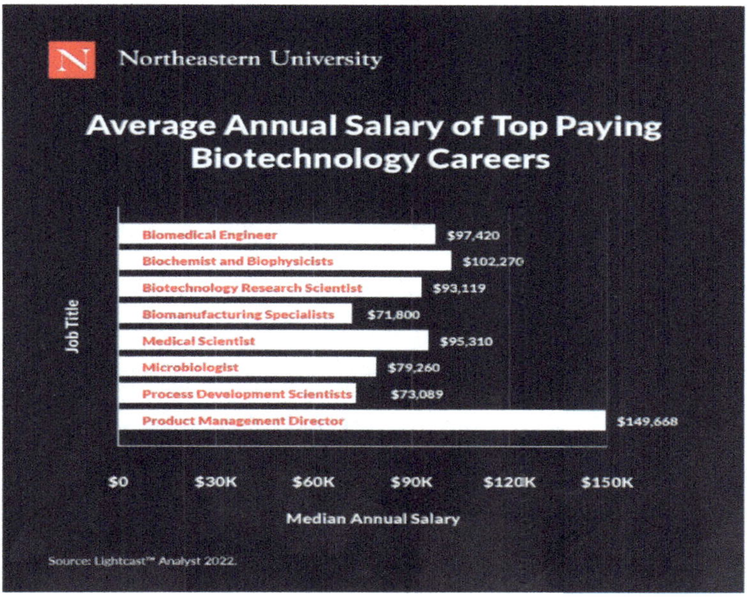

Figure 4: Average Biotech Salary in the US

From Concept to Creation

"When you find your path, you must not be afraid. You need to have sufficient courage to make mistakes"

-Paulo Coehlo, Brida

As with anything, it's very hard to know exactly what the life cycle of a product is from conception to realization. Having this added insight can grant you crucial knowledge to understand the health of the company you're at and even provides an idea of the work load and hurdles too. Let's first try to create a roadmap for a biotech company. (This list isn't exhaustive, and there are certainly gaps in my knowledge surrounding startups formation).

1. *Idea Establishment*

Like with anything valuable in life, a company or person starts with a powerful idea and mission. The very beginnings arise from recognizing that there's an unmet need in the world that should be addressed. To fill this void, the creators can brainstorm on what exactly can resolve this issue, whether it be a new drug or a new type of equipment. This stage requires a lot of personal inquiry and research to get an accurate understanding of the problem to be tackled. Has anyone else started something similar? Is there enough of a market to make this profitable?

2. *Forming a Team*

No problem should be tackled alone, and a good team can be the difference between a life-altering product and a failed concept. The team sets the foundation for everything that comes after, so establishing a group of like-minded professionals with years of expertise is paramount. This includes scientists, academics, legal experts and business professionals. Everyone provides insights and knowledge that the other lacks to set the company up for success.

3. Pre-Seed Stage

With a foundation formed and a team assembled, a solid plan needs to be constructed to most effectively make use of everyone's skills. Crafting a business development plan and demonstration proof of concept are paramount here. A business development plan is a strategy that outlines the goals and actions a company will take to expand its market presence and achieve its long-term goals.

Proof of concept will show to yourself and future investors that what you're trying to develop works. With a biotech company focused on drug development, proof of concept tests the scientific rigor associated with the idea. This can include the hypothesis, an in vitro (think: petri dishes) study to test the drug efficacy, a toxicology report for the safety of the drug and utilizing animal models to see how the products function in animals. There are many other associated steps to demonstrate proof of concept, but its general purpose is to show that the concept has some means of functioning.

4. Seed Stage

With a business development plan and proof of concept in hand, a company is now at a point where it can present its ideas to others and establish itself. Part of this establishment is securing the initial funding that will help the company fund further research and growth. Funding is crucial for any biotech startup. Some sources of funding are: Venture capital firms, angel investors, government funding and incubators/accelerators. I've included brief definitions of each term below:

Angel Investors: Wealthy groups or individuals that provide their own funds to an early stage startup. In exchange, these investors are typically looking for business equity/ownership. Angel investors are typically more risk-tolerant but invest lower amounts of their own money than others.

Venture Capital Investors: Institutional investors that manage a pooled source of money (individual contributors, pension funds etc.). Venture capitalists make use of money from various sources, so they're inherently less risk-tolerant but have larger amounts of money to invest.

Incubators/Accelerators: Both incubators and accelerators are programs that work exactly how you'd imagine – they assist early-stage startups in their development. This assistance can take the form of financial resources, mentorship and networking opportunities. The core difference between the two is that an accelerator is more rigorous in its approach and focuses on accelerating the company's growth, whereas an incubator is more focused on slowly nurturing its growth. It's possible to be in both programs at one point.

Government Funding: This funding is a critical aspect for any early-stage startup. Grants provided by the government (such as the Small Business Innovation Research and the Small Business Technology Transfer programs) are risk free and, unlike Angel Investors, the company doesn't relinquish ownership/equity by accepting it. Financial assistance can help fund research that will define the proof of concept. Other areas of support include technical expertise and a network of qualified professionals.

As you can imagine, a combination of all of these sources of support is critical to establish a successful startup.

5. *Product Development*

In the process of getting a product to market, it's crucial to establish a strong working relationship with regulatory bodies like the Food and Drug Administration (FDA). Background research is conducted to determine exactly what is required for the product being created. The FDA is a government body that determines what drugs and products can make it to market. Building familiarity with their guidelines and procedures can help with having a smooth process.

On top of regulatory upkeep, additional process testing, and optimization becomes crucial. For a drug product, non-human trials are used to test certain characteristics such as how the drug is distributed throughout the body, examining how it's broken down (metabolized) by the body, the acute effects of a single dose and the whether the drug can accomplish its desired effects (i.e. its efficacy).

6. *Series A Funding*

Series A funding is a crucial phase of any biotech company. As mentioned earlier, it's at this stage that Venture Capitalists start to enter the scene for startups to support larger funds. According to DealForma, "The average proceeds per deal in 2023 slightly surpassed those of 2022, with an average of $48.5 million compared to $47.1 million" (Harutyunyan 2024). This money contributes to R&D investments, clinical trials, hiring staff, and overall business expansion.

Another key component is preparing and submitting documentation for future clinical trials. To do this, the biotech company will file an Investigation New Drug (IND) with the FDA to allow it to be used in human clinical trials. Information in the IND includes preclinical/animal trial data showing the efficacy and toxicology of the drug, as well as information about its composition and stability. Failure to have an IND can include termination of research, fines, and prison time.

7. Clinical Trials

Now that there's funding and the staff along with an IND submitted to the FDA, clinical trials are the next step for your product. Clinical trials are conducted on human patients to determine how it responds in people and examine its efficacy. Clinical trials are usually divided into the following:

Phase 1: At this phase, there's a focus on safety and dosage of the drug. The recipients can be healthy donors or patients with the disease in question. This sample size is typically very small (20 – 100 people). The maximum tolerated dose and minimum effective dose are identified to determine exactly how much someone should receive. This phase can last several months to a year.

Phase 2: A larger sample size is selected for this phase (100 – 300 patients). The primary focus of this phase is to assess the efficacy and continued safety of the drug. Adverse effects and toxicity are key indicators that are examined, to ensure the proof of concept is realized. Certain clinical parameters are examined like tumor remission. This phase can also last for several months to a year.

Phase 3: A large-scale sample size is selected for this phase (300 – 2000 patients). Efficacy is confirmed and long-term adverse effects are also noted. An improvement in patient health is compared to other treatment options available. The drug is compared to a placebo to also see that the effects are tangible. This is the longest and most expensive phase, lasting several years.

8. Commercialization

The information from Phase 3 is used to file a New Drug Application (NDA) with the FDA and move the product to commercialization. The manufacturing facility will undergo a robust scale up, including personnel and facilities to assist with reaching the desired capacities. The drug is now at a point where it can be taken to market and become profitable. A market plan for pricing, distribution, partnerships and marketing are developed here. The product will eventually be launched onto the market.

Baby Steps into Biotech

"The journey of a thousand miles begins with a single step."

-

- Lao Tzu

My purpose for writing this book is to provide a comprehensive roadmap that I didn't have. Over time, my passion became about giving this knowledge and insight to others. I've allowed my passion to lay the groundwork for me, and it allowed me to steady my resolve and focus. Entering a new industry with no prior knowledge can be daunting without structure and direction. There's a deficit of resources that can empower you to guide your career. I want to provide you with an accessible way to begin your search and be an inexhaustible resource from which you can pull from.

Getting your bearings

I can't think of a better way to begin your foray into biotechnology than to gather information. I'm sure you do research before you go on vacation or go on a date- you want to know where you're going and what you'll be doing. Likewise, doing the research for your career will set you up for success too. Biotech is an industry very heavily reliant on prior experience. I always wondered why there's such a reliance on experience and how I've come to understand it is that it reduces risk and time needed before an employee can start contributing.

But how do you get experience if you have none? This conundrum plagues the industry and can make it hard to get a foothold. By researching and taking risks on yourself, you can have a targeted approach that will save you time.

Finding a company to join is a daunting task to tackle. Initially I would only look at the job descriptions/posting, but I found this was very short-sighted. It's like wanting to buy a car based entirely on its first appearances without going on a test drive or to a mechanic. It can be done, but I don't think you'd recommend someone to do that.

I'll admit, I've been guilty of blind applying to a job and not following this same thought process. Sometimes it's needed and realistic to do this, but a company you're actively interested in staying at for an extended amount of time should be looked at ahead of time. Some websites which truly helped guide me in my journey were *Biospace.com* and *Bio.org*. Each website provides a strong background on what biotechnology is in order to tell if it's something you'd like or love.

My personal perspectives

The only way to truly grasp what you enjoy is through personal research. I stumbled upon the industry and I'm incredibly happy I did. But seeing how many options there are out there, and with limited time in life to focus on each discipline, having a rough estimate of where you'd like to put your energy is crucial. Don't let tunnel vision blind you to what else surrounds you.

After searching, I found a Quality Assurance role in a Cell Therapy department for Thermo Fisher. In hindsight, a bit of luck and charisma was needed, but I eventually took the role. Many new pieces of information were thrown at me and I felt like a fish out of water. As I continued working, my curiosity into what I was doing began to take hold. I felt a hunger to understand what Cell Therapy was, but I first had to understand Biotechnology on a deeper level.

Let your passion should guide you. I personally felt my innate passions channeled into the work I did. Some days were tougher than others, and imposter syndrome is incredibly tough to overcome. There aren't many black men in the industry and that's always been a big issue in the back of my head. I often found myself questioning my role on teams and pondering my reason for being in the room. Countless times I've questioned my intelligence and work ethic in comparison to others. But I stuck it through, understanding that hard times create strong people.

Look around the world and consider what keeps you up at night. If climate change or pollution keep you up, tackle them with all of your might. Or perhaps you'd love to design the medication or equipment that a hospital may use to better the lives of patients. Odds are there's a company that has focused its energies on that space. But what can you do to actually gauge how good a company is? Some general things I like to keep an eye out for:

- Examine the company's mission, its unique technologies, and key partnerships.
- Look over their website and linkedin accounts in order to get a full picture. When your passions and interests harmonize with your workplace, a greater sense of belonging is forged
- Peer deeply into the company's pipeline, funding/sources of revenue, and its market share. Whilst more difficult with private companies, this information can be found more readily for public companies. I received a lot of help from websites such as *investors.com* and *crunchbase.com*.
- Prioritize your own security. The biotech industry can be volatile depending on the size of the company and what it produces
- Does its technology show promise? Has it received bad press recently that may affect its future funding?

A keen eye can save you from joining a company that you aren't aligned with or makes your development an afterthought. Doing the work beforehand can you save a ton of time and effort in the long run. The exact needs and things you should focus on can really depend on what phase you are in life.

Forging A Career

"You must forge your own path for it to mean anything."

— Rick Riordan, The Lost Hero

College Students

College can be an extremely scary time. It can seem like you should have so much figured out in the span of 4 years with incredibly tough questions to answer: *What do I want to do with my life? What's my purpose? What am I good at?* Each of these can have answers that can change depending on your mood or mental state, and people are evolving all of the time.

There were many times in life when I would say I was going with the flow. I couldn't say any particular direction was strongly dictated by my ambitions but instead by what I was told I should focus on. Thankfully, I'm here to offer some guidance so this doesn't have to be you.

It's very easy to go through the motions, but then you're stuck at a job that might find boring or not making use of your talents. In college, I knew my passion was with the sciences- understanding new concepts and being intellectually stimulated by new processes. However, I also knew that my talents lent themselves more to the creative portions of life, in particular writing and abstract thinking. I intrinsically understood that I had to make a decision on whether I wanted my career to focus on my innate talents or my passions. I simply wanted fulfillment from whatever path

I chose. What I didn't know was that this concept, a life that I feel valued and is worth living, is present in Japanese culture.

By talking with a friend, I was introduced to the Japanese term *Ikigai*. The word "iki" means life and "gai" can be translated to worth or value – literally meaning life worth. It's easy to have a job or a career, but finding fulfillment in all aspects of life requires focus and attention. Below, I've included the general venn diagram of the *Ikigai* concept.

https://www.japan.go.jp/kizuna/2022/03/ikigai_japanese_secret_to_a_joyful_life.html

At first it can seem challenging to decipher each of these concepts, but the benefits are truly worth. If you find yourself uncertain of what to do with your energies, take 5 minutes to ponder each of the core concepts to understand yourself more. As an example, I can give you a quick breakdown of my own *Ikigai*.

My *Ikigai*:

a. What am I good at?
    i. Writing
    ii. Speaking
    iii. Learning and understanding people and concepts
b. What can you be paid for?
    i. Writing books
    ii. Public Speeches
    iii. Consulting
    iv. Working in a science focused area
c. What The world needs?
    i. More black men and creatives in STEM
    ii. Novel approaches to problems

iii. Compassion and understanding

d. What you love?

i. Learning and self-improvement

ii. Helping others

iii. Expressing myself via powerpoints, speeches or writing

iv. Making the world slightly better than it was before

So taking all of this, how does everything overlap?

a. **Profession**: I'm great at understanding concepts and speaking. STEM has a great need of someone who enjoys to understand and breakdown technical concepts to others. I've seen this manifest largely in technical training roles I've taken on in my time in Biotech.

b. **Passion**: I truly love helping others and making the world slightly better for others. I'm great at speaking and writing. My passion manifests itself in either mentorship or presentations I enjoy giving to others.

c. **Mission**: The world is in dire need of different perspectives brought to STEM. These different perspectives are the bedrock upon which the iterative nature of science is built. I love helping others and expressing my creativity while doing it. This book itself is a testimony to my mission in action.

d. **Vocation**: The world needs novel approaches to problems and compassionate, creative approaches to achieving them. I can be compensated by doing this in many ways, whether it be personal consultations or even panels that can help students understand Biotech.

It took me several years to really understand how everything can piece together, some of it only manifesting in the year this book is being written (2024). Experience is really the best teacher, and some of the best ways to gain it is via an internship or a mentorship. Odds are your college offers some form of internship/mentorship guidance. You can also take it into your own hands and check out companies on LinkedIn to see openings.

Putting Pen to Paper

The biotech industry is a very niche and specialized field. On the one hand this is great: once you've attained a foothold, your experience makes you invaluable. On the other hand, actually attaining this experience becomes difficult. The time in college is spent learning new and challenging concepts, completing assignments within a given due date, honing your lab skills and collaborating with other students. All of these skills are used regularly in the workplace. Even though there are roles that can cater to people of all academic backgrounds, certain majors really see huge returns on investment.

The degree you choose can really make a difference in how prepared you are for a career in Biotech. The degree offers good exposure to the necessary knowledge, skillsets and principles that can apply directly to the job you're striving for. Some common bachelor's degrees present within industry include Biochemistry, Microbiology, and Chemical Engineering. When writing on the Biochemistry major, CollegeConsensus.com notes "Many biochemistry majors work in medical laboratories or pharmaceutical companies. Often, biochemistry majors enroll in medical school, pharmacy programs, or doctorate programs" (CollegeConsensus 2024).

A degree that offers insight into the human body and cellular biology is irreplaceable. I've come across management level employees that have had undergraduate degrees in biochemistry or chemical engineering. Extensive experience in a lab made a crucial difference when I was applying to jobs. Working in a fume hood, following and executing experimental protocols, and using common lab equipment (scales, micro pipettes and water baths) made a difference when I applied to entry level jobs.

Some jobs to help you begin your search include:

- Quality Control Technician
- Quality Assurance Specialist
- Process Development Associate
- Manufacturing Assistant
- Animal Technician
- Clinical Research Associate

However, don't let this dissuade you if you don't have a degree in the life sciences. Every major calls upon different mental processes and techniques that allow you to approach problems with a unique perspective. The life sciences can give a great leg up in understanding core concepts associated with the company, but it's not impossible to gather that information on your own.

There's a saying that I'm sure we've all heard, "It's not what you know, it's who you know". Coming out of college, I had ZERO academic references I could really lean on. I depended entirely on my ability to connect and sell myself to people. I love speaking with people and connecting with them. It just took me leaving college to really make the most of it. While I was blessed in acquiring a job so quickly (which speaks to the power of having an undergrad degree in Biochemistry) I don't want you to rely on blessings and luck.

Be active and present in your classes. Even if it's just one class, make sure your professor knows you. I went to a school of 30,000+ students, so being front and center was critical. Hopefully your classes have fewer students making it easier to have your professors know you. Go the extra mile and join clubs or organizations while you're a freshman or sophomore which can go a long way to getting you looped into potential opportunities. Take the time to really connect and leave an impression of yourself with them.

College Grads

When I first started in industry, I was fresh out of college with a biochemistry degree. Unfortunately, I did a terrible job in preparing myself for the future. I didn't have any prospects lined up or companies lining up to get into. I worked as a security guard for a few months just so I'd be able to pay my bills. In hindsight, I was a mess.

I remember waking up one day and thinking "I didn't go to school for this. There's no way I suffered and struggled to have a job that didn't make use of my degree's skill sets." I decided to set my sights on something more suited to my goals. I found a contract role in an industry I had no idea about- Cell Therapy. It seemed to make the most of my skills and the knowledge I gained. Suffice it to say this was one of the best decisions of my life. I've always been one to jump into a new opportunity and make the most of it.

The first job without experience is an uphill battle. The key to succeeding is playing to your strengths. I'm a great interviewer thankfully and, coupled with a bit of luck, I secured the role. When I initially started working at Thermo Fisher, I didn't have much guidance on what to do with myself. My original contract was 6 months, and I was more than ready to leave for a new opportunity. To this day, I'm thankful that I had a life altering conversation with a coworker.

She was a senior member on the team and noticed I had no intention of staying. She turned around in her seat in our small office and pointedly asked me: "Do you feel like you're using what you learned in school?" This upfront question was something I should've asked myself, but I wholeheartedly believe she was put in my life to enable me to reflect and ponder this. My response to her was "No, not at all." We began to have an hour- long conversation with each other about life, our purpose, and what it meant to be a minority in a field that didn't have many other black people.

Full-time positions aren't always available so there's no harm in taking a contract role. It will provide you with invaluable experience and connections that will carry you through to even better opportunities in the long run.

A common question I've had to ask was whether an additional degree was worth it or if my years of experience would be able to carry me. Anecdotally, I ran into people across the board who had very advanced degrees (Masters, PhD, MD), a bachelor's degree or no degree at all. What I found was really enlightening for me. An advanced degree plus years of experience was the game changer, where you could really get a voice in the room and have strong leverage in many roles.

A bachelor's degree with many years of experience (10+) will do wonders, especially when you have great references and can speak to your accomplishments. However, that lack of a more advanced degree can create a glass ceiling that can be harder to overcome. Once you've settled into a role, take some time to see how you'd like to develop your career (*Ikigai*).

Develop Your Learning Style

"I am tomorrow, or some future day, what I establish today. I am today what I established yesterday or some previous day."

— James Joyce

Just as important as learning is knowing *how* to learn. Knowing how you learn best can really make the difference with anything you do. To be clear, people learn differently at specific ages, points in life and environments. How you learn as a student in college can definitely change by the time you're a working professional. My goal by highlighting the different learning styles isn't to constrain you to one or the other, but to give you the tools to know exactly what appeals to you in the moment.

There are typically 4 buckets that people will lean more towards in terms of their preferred learning strategies: Visual, Auditory, Read/Write, and Kinesthetic.

Visual

As the name implies, visual learner respond extremely well to visual aids. This can include pictures, graphs, or videos that can help them retain the information well.

Tips: If you're taking notes, consider using symbols or icons to represent what you're writing about. For example, every time you're writing about a plant maybe you draw it. Or, if there are a lot of numbers consider putting into a graph to show their relationship.

Auditory

An auditory learners retain information best when they hear it aloud. Group discussions, lectures or videos work extremely well. I personally associate most closely with this group.

Tips: Engaging in group discussions is crucial to the learning process. Not only that but reading items aloud can work wonders.

Read/Write

These individuals can learn best by reading and writing the information provided for them. So, if they're given a textbook or documents at work they can learn the information they need by reading the information, digesting it, and writing it again.

Tips: Develop a powerful note taking system. Not only should they be unique, you want to be able to review it continuously so the information can stick.

Kinesthetic

Finally, the kinesthetic learner does best when they're able to physically put into action what they learned. Their learning style is very much tied to all of their senses. This is particularly useful when a lab is available that requires them to put what they learned into practice.

Tips: Attend any labs available to you. However, if a lab isn't available, try creating an example of each concept you just learned. Making the example more personal can make it easier to retain and recall.

While it'd be nice to neatly fit into any of these examples, it's rarely that cut and dry. Sometimes there's an overlap between them that contribute to your overall learning style. Maybe you enjoy physically drawing diagrams on a whiteboard (kinesthetic and visual) or maybe you do great speaking and explaining a graph you created (auditory and visual). As I mentioned before, take the time to experiment and see what dynamic keeps you engaged and excited.

Getting a Good Start

"Even the greatest was once a beginner. Don't be afraid to that first step."

-Muhammad Ali

Understanding your company and your role in the first few years can be both exciting and scaring. By getting a solid first start, your pathway for growth will become more defined and easier to execute. For my first year in industry, I felt like a fish out of water. Companies can be large or complex. It can be tough to understand each department, its overall role for the company's mission, what skills are developed and the means to transition into that area.

Prior to understanding the company, I think a better idea is understanding yourself. What is it that you hope to accomplish by being there? What is considered successyou're your role? There are some important milestones in your first year that, if you focus in on them, can help you be successful in the first year. Namely, setting strong goals for your first 30, 60 and 90 days.

30 Days

Kicking off your new job, the first 30 days are crucial to how you start. A good start is important to your perception of yourself, how others view you and your ability to have an impact. Some things to make note of during this time:

1. Ask questions: You want to make sure you're understanding the exact intricacies of your role and the process you're working on. Make time to ask questions

to members on your team or even other teams. Even more importantly, find ways to find time on your coworkers' calendars. This can help with introductions and familiarity with the company. Examples of great questions to ask are:

   I. "Can you connect me with any teams or individuals I'll be collaborating with?"
   II. "How can I best contribute to the team's goals over the next few months?"
   III. "Are there any key processes or workflows I should be aware of?"

2. Set up check-ins with your manager: Your manager is the most important person for you. They are a resource for you to tap into to get a foundation for what more you can do, what's expected of you, and to see where you can help. A great question to ask is, "Where can I help make your life easier?" Establish a positive relationship can be the difference between enjoying your job and feeling lost.
3. Identify knowledge gaps: An important aspect of growth is having the humility to realize that there's always more to learn. If your company has a process description of its major products/projects, make sure to read that over. Test yourself to see if you understand everything. For me, if I can't explain it, I don't know it. This is also an

excellent opportunity to connect with people from other teams who are SMEs on certain portions of your company's process. As you'll see in the next chapter, gathering insights from other teams can be insightful and empowering for your development.

60 Days

The 60 days are where you can create an action plan around what you've learned in the first 30 days. This action plan becomes a powerful tool for you to direct your energies and efforts. Consider the difference between running water and when you've attached a hose to harness and focus the water into something more powerful. Some tips for this phase:

1. **Create SMART goals:** SMART is an acronym for: Specific, Measurable, Attainable, Relevant, and Timely. By ordering your action plan with SMART goals, it will be something that's focused, powerful, and aligned with your intentions.
2. **Create an action list:** The benefit of an action plan means breaking up a desired goal into smaller objectives that will eventually get you to that point. By doing this, "you can make the final goal seem less overwhelming and you can move forward in an organized manner" (Herrity 2024). Kind of like trying to carry an entire

couch vs. taking out the cushions and then moving the couch.

90 Days

At 90 days, a new level of comfort and development should be taking place. You want to begin becoming autonomous and taking on tasks on your own, being willing to volunteer yourself to help. This is an excellent point to take the initiative on a project that you're interested in. Most importantly check in with your manager to assess where you're strongest and weakest.

Crafting A Niche

"We are all apprentices in a craft where no one ever becomes a master."

— Ernest Hemingway, The Wild Years

Understanding the biotech space and what it is you're doing will go a long way to establishing a connection to your work. I always find it easiest to envision its core departments as a car: Manufacturing is the engine, Quality Control is the brake, Quality Assurance is the steering wheel, R&D is the transmission and Supply Chain is the tire. When they're all pieced together, you have one machine moving towards tackling a specific disease. If you don't know anything about cars that's ok, I'll break it down for you below.

Manufacturing

Going by our metaphor, Manufacturing is very much the engine of the organization. It's what drives the organization's efforts, producing the pharmaceutical products to help save lives. Without the efforts of Manufacturing generating results and data, the biotech industry wouldn't be able to function.

The Manufacturing department relies on skills such as technical expertise in bioprocessing, including use of a micropipette, syringes, and centrifuges. They follow Good Manufacturing Practices (GMP) to ensure products are safe and consistent. This department focuses on processes like incubation and purification, adhering to strict quality standards. Team members need to be proficient in operating specialized equipment and following GMP guidelines for quality and safety. They should also possess problem-solving skills to troubleshoot issues that may arise during production.

Photo by Luca Hooijer on Unsplash

Quality Control (QC)

Quality Control acts as the brakes for the biotech journey and helps to ensure the safety of the products much like the brakes on a car. The department checks and verifies the quality and safety of biotech products, preventing potential issues and ensuring smooth progress. Important tests like Gram Staining are conducted to test for any bacterial contamination which could harm already sick patients.

QC personnel must be skilled in various analytical techniques. They use techniques like High-Performance Liquid Chromatography (HPLC) and Mass Spectrometry to verify product purity and safety, while maintaining compliance with Good Laboratory Practices (GLP). They need strong attention to detail and the ability to follow protocols precisely. Critical thinking is essential for data analysis and interpreting results to ensure product quality.

Quality Assurance (QA)

Every vehicle needs a steering system and that's where Quality Assurance comes into play. QA guides the company on the right path, making sure it adheres to the regulations and procedures of the FDA and the company's policies. Much like how steering ensures the car's direction, this department ensures compliance with regulatory guidelines. QA keeps every company from veering off of a safe road and over a cliff.

Quality Assurance professionals require a deep understanding of regulatory requirements and compliance with GMP, GDP, and GCP standards. They monitor compliance with Good Documentation Practices (GDP), GMP, and Good Clinical Practices (GCP). This department acts as the gatekeeper, preventing product release if it doesn't meet required standards. Communication skills are crucial for coordinating with other departments to ensure that processes and documentation meet the necessary guidelines. They should also be adept at risk management and auditing.

Photo by Laura Gariglio on Unsplcsh

Research and Development (R&D)

R&D is like the transmission of the biotech organization and assists with adaptation and progress. The transmission of a car facilitates change between gears for optimal performance. R&D allows the organization to tackle new challenges with new technologies and tactics. It drives innovation by developing new products and improving existing ones, giving a company the ability to adapt to new roadblocks.

The R&D department relies on scientific and technical skills, including molecular biology, genetics, and biotechnology techniques. Research scientists need to have strong problem-solving abilities to overcome challenges in product development. They should also be proficient in data analysis and have the creativity to innovate in the field. They work on cutting-edge technologies, conduct clinical trials, and constantly seek ways to enhance drug effectiveness and safety.

Photo by Maxim Hopman on Unsplash

Supply Chain

Supply Chain acts as the tires of the biotech industry, connecting the organization to the necessary resources. For example, the material needed byManufacturing or R&D to conduct their work will come from Supply Chain. Just as wheels and tires provide stability and traction, supply chain ensures the smooth flow of materials and products, linking the company to its vital components.

Supply Chain professionals need organizational and logistical skills to manage the procurement, transportation, and distribution of materials and products. They should have a solid understanding of GMP and GDP regulations to ensure product integrity during transit and storage. Communication and negotiation skills are vital when working with suppliers and distributors to maintain a functioning supply chain. They coordinate production schedules, ensure efficient distribution, and maintain proper storage conditions to meet GMP requirements. This department plays a vital role in ensuring timely access to life-saving medications.

Photo by Hoyoun Lee on Unsplash

Depending on the size of your company and your role, cross functional support is always present. Take advantage of it by talking to new people. At first glance, it would be easy to think each department is completely distinct from one another. However, there is a copious amount of work that depends on cross functional support. A great example would be with MFG and QC. During a manufacturing process, the ultimate goal would be to create a drug product. However, with lots of variability between different processing days, the culture being manipulated changes. The MFG team will take different samples. QC will take these samples and use them to conduct an analysis of the sample: checking for characteristics like purity, concentration and for any contaminants. These tests from QC enable the release of the drug product, concluding the MFG process. This overlap generates a ton of teamwork and coordination between members of both teams, creating a fully functional vehicle for product development!

Photo by Kartik Bhattacharjee on Unsplash

Seeing Is Believing

"Faith is taking the first step even when you can't see the whole staircase."

— Martin Luther King Jr.

I would always daydream as a kid. In class, I would find myself looking out the window and envisioning some obscure reality. The same would happen on the bus rides home. I truly thought there was something wrong with me; that I had an inability to focus and pay attention. But I realized that daydreaming was my superpower that helped strengthen my vision. When speaking on the impact of daydreaming on creativity, it's been noted that "But when people's minds wandered in more fantastical ways…or in ways that seemed particularly meaningful to them, they tended to have more creative ideas and feel more inspired at the end of the day, too" (Suttie 2021). Allowing your mind to be open to more increases what you're able to understand as possible to achieve. Your vision sharpens and your field of view widens.

Vision will guide you and keep you from getting lost in life. It's the thread that connects our past, present and future, helping to keep us from getting derailed by the commonalities of life. I implore you to allow yourself the freedom to see where you fit into this world. To see what definite major purpose you have that no other person can hope to replicate in the exact same way. What can YOU do that you're so confident in that nobody can take away from you?

"But Tejon, how can I know what to do if I have no idea what's possible?" This right here is something I struggled with for a long time as well. How is it possible to know what role to envision myself in if my vision is clouded with possibilities and uncertainty? It's at this point that I would suggest surveying the land. Having a reference point to compare against is invaluable.

Take a deep look within yourself and consider what you like. Maybe you enjoy talking to people, or you like being in a lab. Maybe sales attracts you a lot or simply training others is your calling. Biotech is an incredibly diverse industry that can cater to many skillsets. Look at someone in a role you want. Find out how they got to that position (*Informational Interviews*). How long did it take? What prior roles did they have to push through to get there? Having this reference will allow you the ability to plot a path forward. Your vision, in conjunction with your Ikigai, will keep you in tune with yourself.

Definite Major Purpose

A short time ago, during the COVID-19 Pandemic, I began searching for stimulating reads and audiobooks. One such audiobook I came across was Napoleon Hill's *Keys to Success: 17 Principles of Personal Achievement(Think and Grow Rich)*. Hill's book is one of the most widely sold self-help books since its release in 1937. The book expounds upon several core topics and mentalities that successful people adopt. One mentality is Positive Mental Attitude (PMA) and I'll touch on this in a later chapter. Another key concept he brings up is this idea of a Definite Major Purpose (DMP).

A DMP can be seen as a goal critical to your life's goals that you can't go without it. When Hill describes DMP he points out, "If you know what you want from life, if you are determined to get it to the point that it becomes an obsession, and you back that obsession with continuous effort and sound planning, you have awakened definiteness of purpose" (Hill 2011). Your plans and efforts in life should align with your DMP, and your vision will enable you to understand the different opportunities set before you.

We talked earlier of your Ikigai, where you can develop a comprehensive view of fulfillment for yourself. Lumped into your Ikigai should be your DMP, most closely relating to your mission. Consider the impact and legacy you want to leave behind in the world. It can be something that keeps you up at night and you can't figure out why nobody else has done it yet. No matter what it is, make sure YOU believe it wholeheartedly.

Have A Little Faith

Having vision and purpose are amazing on the days where you're succeeding at your goals and seeing progress in life. But I think we all know that that's just not how life works. More often than not, we'll have our patience, resilience, and tenacity tested, and our resolve shaken. Doubt creeps in to our hearts, making us susceptible other intrusive thoughts. How do we get back on the horse after we've fallen off?

This is where faith comes in. To win and overcome requires a strong mental fortitude to continuously overcome hurdles and setbacks. On a day to day basis, "Faith gives a strong belief that we'll win in the end and removes the thoughts that make us afraid, such as thinking about worst-case scenarios" (Alive Foundation 2022). Fear and doubt can limit us from reaching our potential, and can cloud our vision from focusing on our DMP. Faith is a powerful remedy to these toxic thoughts.

By definition, faith can be described as complete trust in someone or something, typically in religious doctrines and a deity. As a Christian, my faith is tied directly to the utmost trust and belief in God's promises to humanity. On my toughest days, I look to Him and understand that he's gotten through some extremely tough moments in my life. Whether that's the loss of both of my parents and grandparents, or the struggles I faced in college, His plan included me going through those events and become the man I am today. My faith becomes a shield to better enable me to overcome problems, stay true to my principles, and be a better man every day.

Even if you don't subscribe to a religion, having faith plays an important role in how you move through life. To some degree, faith is trust that something will work properly. Think about the last time you were on a plane or a train – yes, science holds up that the physics behind these machines should work. But you have to trust that it's been properly maintained or inspected for any deficiencies. Truly, faith makes it easier to function everyday, "erases doubt and the notion that things are doomed to fail" (Alive Foundation 2022). Unless you expect to see the reports indicating that all of these responsibilities were performed by trustworthy and skilled people, you're operating on faith that they will work. By combining faith, DMP, and vision you're primed to set along a career path that's intentional and powerful.

Everyone Needs Support

"Close friends are truly life's treasures. Sometimes they know us better than we know ourselves. With gentle honesty, they are there to guide and support us."

- Vincent Van Gogh

Years ago while I was in college, I distinctly remember having a panic attack. I looked around in my life and I felt so selfish and isolated; I didn't give much thought to what a relationship meant to me. If something happened to me, would anyone care? Would I leave a positive mark on anyone's life? Looking back now, I can see how foolish that was, and how I let doubt and fear cloud my vision. However, it did have one positive benefit: It helped me understand that every relationship requires an investment of time, support, and understanding. This is the basis of a support system.

A strong support system begins with family and close friends: people whose trust you've helped to develop and cultivate. When speaking on the importance of friendship, Mayo Clinic points out that friends can "Increase your sense of belonging and purpose, boost your happiness [and] improve your self-confidence and self-worth" (Mayo Clinic 2022). In moments of extreme duress or conflict, your support system can literally act as life support for you, enabling you to endure hardships better. One small conversation can have a magical effect on your mindset and open your eyes to new avenues of thought. By taking their experiences and incorporating them with your own, you'll be able to gain a greater vantage point to be able to better choose your path.

I always struggled with opening up and allowing others into my safe space. I always figured I could handle everything on my own, but I couldn't have been more wrong. By closing myself off, I attempted to handle stress by myself and missed out on such key advice that it made my life worse off. I was inadvertently sabotaging my own progress. I learned how to build a support system with other friends for specific purposes. I have friends who can offer fantastic financial advice, can instruct me on career do's and don'ts and even relationship advice. Cultivate the friendships you need.

Just as important as building a support system is nurturing it and keeping it healthy. Imagine you're a gardener- simply planting the seeds isn't enough. Your plants must be carefully tended to provide the nutrients for them to grow. When speaking on friendship, Mayo Clinic points out "Developing and maintaining healthy friendships involves give-and-take… It's as important for you to be a good friend as it is to surround yourself with good friends" (Mayo Clinic 2022). A gardener that forgets to water their plants or place them in the sun can't be surprised when they fail to grow.

Professional Relationships

Of course, it'd be short sighted to only incorporate loved ones and close friends. Like your personal support system, your work network is incredibly important too. The people you work with or meet become integral to your career growth and development. Their guidance and experiences can be great fountains of knowledge to point you in the right directions.

First let's start with the elephant in the room: what *is* networking? I'm sure you've heard it tossed around, but what exactly is it and what is its purpose? Networking is the cultivation and development of beneficial relationships with others in Biotech. Whether at the same level, senior to you or junior to you, by developing relationships with others, you're able to enhance your knowledge and establish meaningful relationships. This knowledge can be from the experience that people with 20+ years over you can share about their paths to success or failures they learned from. It can also be fresh eyes and new perspectives offered by someone just getting started. A network can be means for you to grow and can mean new opportunities further down the line. Some tips that can help you in your endeavors to establish and nurture a network:

Join a professional organization- A professional organization can be pivotal to how you building out your network. These orgs are amazing for meetings professionals in your field, finding new job opportunities, and gaining access to virtual or in-person events. Biotech is such a relatively small industry that the professionals in these orgs have likely worked at or known someone in a place you'd love to be at. Some examples include the American Biotechnology Association, BIOCOM California and the previously mentioned Biotechnology Innovation Organization.

Ask for a mentor- Mentorship is a powerful tool in any industry that allows the experience of one to greatly shape the growth of another. In many ways, a mentor acts as a steward for your own professional development. For example, a mentor "Mentors encourage and enable another person's professional or personal development... the mentor can connect [mentees] to potential opportunities or individuals who can help them" (Birt 2023). The important aspect is finding someone you know and trust enough to establish this relationship. Some good places to start are potential labs or classes you've been in, or former coworkers/bosses that can be a fountain of knowledge.

Become a connector – I can speak strongly about being a connector. This is a skillset I employ on my day-to-day, whether out with friends or at church. A connector is exactly as it sounds: a person that connects two individuals with similar interests that otherwise wouldn't know each other. In fact, “Introducing two people who could mutually benefit from each other can be an effective way to amplify your own value creation. Be generous with your contacts and your network may expand exponentially” (Sklar 2024). There’s strength in numbers, and having two people who can recount a positive interaction due to your actions goes a long way to developing how others see and remember you.

Each of these qualities build upon each other and require intentionality on your part to make sure they’re given the proper nourishment. A dependable person is someone you know you’re able to call in your time of need and entrust with your time. A relationship founded on trust enables you to speak on your flaws, weaknesses and vulnerability that emboldens each party to go above and beyond for each other.

Persevere... and Then Persevere Some More

"You don't have to become something you're not to be better than you were."

- Sidney Poitier, *Measure of a Man*

The famous Greek philosopher Aristotle once said "He who has overcome his fears will truly be free." It is by overcoming fears of ourselves and our limitations that we can become something greater. Certain levels of career success necessitate a willingness to take risks outside of our comfort zones. Whether this be in the form of striking up a conversation with a member of a separate functional group or being bold in your desires for a promotion, by overcoming yourself you can reach heights unprecedented.

In chemistry, there is a term called a "limiting reagent" during a reaction. No matter how much of the other material you may have, if you run out of that limiting reagent, the reaction will cease to occur. I've talked a lot about getting others to support you and taking steps to support yourself. In my humble opinion, none of this will matter if you allow yourself to be the limited factor. Put another way: you can have the nicest car, with the newest tires and best seats. None of that will matter if you can't afford to put gas into the tank or air in the tires.

We all have limitations. I, for example, hate making mistakes. My fear of making mistakes prevents me from trying new things so I don't look foolish or incompetent. When speaking on the importance of stepping out of our comfort zone, Deniz Kayadelen highlights "Stepping out of your comfort zone makes you more creative, decreases your own limiting beliefs and enables you to use new ways of thinking and acting" (Kayadelen 2020). Our brains want us to be safe and comfortable on a day to day basis. This is why intentionality in our actions matters so much. What I tend to find when I step outside of this comfort bubble is that the results are much better than I could've imagined, and I learn something new in the process.

Taking a different approach

Let's take a second to peer into a few of Kayadelen's pointers about overcoming this fear. What we'll find is that several of her points tie neatly into concepts we've already touched on in previous chapters.

1. **Give up control and trust the process**
   There are so many things in life outside of our control that thinking we could control them would drive us mad. Kayadelen points out "Do everything you can do, believe in what you do and want, be also be open to letting things go and trusting the process" (Kayadelen 2020). This sounds very much so like our topic on faith in previous chapters. Failures will occur at any job, especially in one as technical as Biotech. Sometimes they have to occur for us to best learn. This will take a lot of the pressure off of you when making critical

decisions. In fact, the *Serenity* prayer highlights this perfectly, "God grant me the serenity to accept the things I cannot change; courage to change the things I can; and wisdom to know the difference". Even if you don't subscribe to the Christian faith, the context of the prayers calls for us as agents of our lives to have courage and strength in our actions.

2. **Follow your passion**

   By following our passions, we can make the most of our talents and energies. Kayadelen alludes to this when she says, "Identify activities that energize you and make you feel happier and stronger in your overall life" (Kayadelen 2020). This happiness will make overcoming our limits all the more possible. In Biotech, there are many processes that will be new at one point or another and will require a passion to drive you. This ties in perfectly with *Ikigai* and a need to identify what most aligns with our desires.

3. **Think big and start with small steps**

   Previously, we talked about the importance of goal setting and vision. By having a grand vision that's followed by SMART goals, you can accomplish amazing goals. Kayadelen touches on this as well, stating "All big things start with small realistic steps, which set the base for preparing yourself for bigger steps. There are no limits to what you can accomplish- except the limits you place on your own thinking" (Kayadelen 2020). You're absolutely capable of amazing things when you put your mind to it.

Ask yourself: "What areas am I struggling with? What am I afraid of trying?" Be honest with yourself and once you hone in on something, take the steps to do it. And once you've done it, do it some more. By strengthening your weaknesses, your limiting reagent becomes less of a hindrance and your success can materialize more easily.

Maintain a positive mental attitude

Earlier, when speaking on Hill's *Keys to Success*, I mentioned that one core tenet he brought up was a Positive Mental Attitude (PMA). A definition provided by Courtney Ackerman puts it as "Having a positive mindset means making positive thinking a habit, continually searching for the silver lining and making the best out of any situation you find yourself in" (Ackerman 2024). It's impossible to overstate how important your mentality can affect your actions and growth. Every small hurdle becomes like a mountain when trying to overcome your own mindset. It's hard enough when the world is moving against you and making it hard to succeed. If you aren't in your own corner, you're fighting yourself.

A negative mentality can cloud your vision. As mentioned, a PMA helps you see the silver lining in all things- it can literally affect whether you view something as a burden or an opportunity. A level of optimism grants you the clarity to see how certain decisions can be life changing. For example, by taking on a task that wasn't within your scope you not only learn something new, but allow others to see you in different capacities which can open doors for you. Your mentality can affect how others wish to work with you, how well you learn new skills and even your physical health!

I do want to emphasize that I don't expect anyone to be happy all of the time- I'm not happy all of the time! However, optimism doesn't necessitate you turning a blind eye and seeing every situation as good or happy. In fact, Ackerman points this out when they say "Developing the right thoughts is not about being constantly happy or cheerful, and it's not about ignoring anything negative or unpleasant in your life. It's about incorporating both the positive and negative into your perspective and choosing to still be generally optimistic" (Ackerman 2024). It's important to take the good with the bad and understand that they each hold their own level of value. How could we evaluate good without knowing what's bad?

Broaden Your Knowledge

"Reading furnishes the mind only with materials of knowledge; it is thinking that makes what we read ours."

— John Locke

Expanding your industry knowledge enriches you, setting you apart from the competition and sharpening your mind. Staying informed is valuable, but the key first step is to engage with someone more experienced. For me, that meant finding a mentor. My mentor, Ken, started as my boss and quickly became a trusted guide and friend, sharing invaluable insights that accelerated my growth

Staying updated on biotech news is challenging, but highly rewarding. With countless companies achieving varying levels of success globally, tracking everything can be overwhelming. However, the payoff is deep industry knowledge, helping you identify in-demand techniques and widely adopted technologies. This insight will put you ahead of the curve in learning key skills and even uncovering promising investment opportunities based on emerging breakthroughs.

Combining the wisdom and experiences of others along with the free knowledge provided by biotech organizations can equip you with the tools for success. Nobody can achieve success alone and it's always good to shore up support and assistance whenever possible.

A resource is any asset that can be drawn upon to function more effectively. This can be a person, a news article, or even a book. Part of strengthening your position includes utilizing pre-established resources that others have taken the time to piece together. Some great resources that helped me include *Biospace* and *Fiercebiotech*, organizations with easily accessible news in biotech, including companies, stocks, acquisitions and layoffs.

Upskilling

In an ever-growing world with technology advancing at a rapid pace, there's a pressing need to differentiate one's skillset from others. Workers HAVE to continuously grow and develop in order to remain relevant in the workplace. This is where upskilling comes into play. Upskilling can be described as the process of training and development that allows for the expansion of one's skillsets. Forbes put it best when it said that "To ensure future career success, professionals will need to be proactive and adaptable to stay relevant and thrive in their careers" (Dennison 2023). Upskilling allows for a dynamic and powerful career and positions you for the next steps in your career. But what does that look like?

Part of the equation is grasping the actual value of upskilling. A few examples of its value can be seen below:

1. **Increased Job Security**: By adding new skills to your repertoire, you become more valuable to your boss and company.
2. **Higher Earning Potential**: Additional skills intrinsically give you more leverage when acquiring a new job, allowing you to command higher compensation packages.
3. **Adaptability**: The new skills you learn better prepare you for changes in your industry and can give you the capacity to pivot to new roles as they become available.

Whether you're in Biotech or a completely separate industry, the value of upskilling sells itself. It's not easy and requires constant effort, but the reward will be worth it. So, what are some skills you should tackle to be set up for success?

1. **Current Good Manufacturing Practices** (cGMP) form the foundation of the biotech industry, particularly in manufacturing. cGMP ensures that medications are produced safely and consistently, guaranteeing that the final product is safe for patient use. Key roles in this process include positions in Manufacturing and Quality Assurance, which are critical in upholding these standards and ensuring product safety.

2. **Flow cytometry** is an in-demand skill that allows scientists to count, analyze, and sort cells in a sample. Using machines like the **BD FACSCanto** or **Beckman Coulter CytoFLEX**, this technique employs lasers, dyes, and markers to identify cell population types and determine their size. Flow cytometry is widely used in both Quality Control (QC) and Research & Development (R&D), making roles in these departments ideal for gaining hands-on experience and deeper knowledge of this technique.
3. **Design of Experiment** (DoE) is a methodology used in biotech for systematically planning and conducting experiments to explore the relationships between factors affecting a process and the output of that process. It enables scientists and engineers to optimize experiments by identifying key variables and their interactions while minimizing resource use. DoE is frequently applied in both Research & Development (R&D) and process optimization, making roles in these areas essential for those seeking to master this analytical approach and contribute to improving product quality and process efficiency.

Be A Team Player

While this may seem like a generic statement on a resume, strong interpersonal skills are truly transformative in how you approach problems and collaborate with others. No matter how technically skilled you are, success in any industry depends on how well you work with others. In such a tightly interconnected industry, building positive relationships and a strong reputation will often take you further than technical knowledge alone.

Sacrifice is a key trait in a successful team environment. This could mean volunteering for extra tasks, taking on more responsibilities, or helping other teams with their projects. Putting the team's needs ahead of your own is admirable, but balance is important. Do your best not to overcommit yourself to the point of burnout. Prioritize your own well-being to maintain performance, while contributing effectively to your team's success.

Parting Thoughts

"I learnt that the only way to get thing done is to start to do it, then keep on doing it, and finally you'll finish it, even if in the beginning you think can't do it at all"

– Langston Hughes, The Big Sea

We did it! We've made it to the end, and in doing so, have demonstrated the most crucial quality a person can have: discipline. Consistency and discipline will be the key to following and executing the principles laid out here. The best learners take what they're given and enhance it. I'm confident you'll think of things beyond what I've shared in this book. This is merely a framework for you to build on as you continue your journey. Despite this being the end of our journey (for now), I do want to highlight some parting words to you.

First, your resilience will take you further than any other aspect you may have. Any journey worth traveling is rife with hurdles that will undoubtedly test your capacity to overcome them. Some of the most transformative experiences require us to feel shame, doubt and worry before we can strengthen ourselves to them. Remain resolute in your ambitions and don't allow the present to dissuade you of what the future may hold.

Learning is one of life's joys and it should never stop, even with the ending of this book. Much like you, I anticipate walking away from this book more fulfilled and complete than I was when I first began. Also like you, I will acknowledge there will be new experiences and knowledge that I will gain that won't make it in this book. I implore you to always find new avenues of growth and development, gaining new knowledge, skills and insights along the way.

Progress is not guaranteed and should be celebrated at every opportunity, no matter how small the victory. The reality of your accomplishments must be acknowledged to provide them with an added layer of depth. Pride is a tricky thing: too much and you run the risk of false attributing your successes. Too little and you run the risk of displaying false humility, preventing you from appreciating your skills and talents. A healthy balance can let you remain humble while still appreciating the blessings bestowed upon you.

Along the way, you may question your capabilities or compare yourself to others. I've struggled with my place, not just in this industry, but in the world. While I can't give you a reason for being, I can assure you that everyone has one, even if they haven't discovered it yet. Be patient with yourself. The brightest stars may shine twice as brightly, but they burn for half as long.

Ascent

I climb and look above

my vision wavers, my resolve falters

Pain echoes through my body like a shout in a cave

Panic pours from my pores

My heart beats like a drum

I shake from the sight

My eyes shift down the mountain

The clouds passing under my feet

The way down further than up

Serenity falls over me like a blanket

Power returns in waves

My gaze changes, and courage swells

- *J. D. Rae*

WORKS CITED

Ackerman , C. E. (2024, July 23). *Positive mindset: How to develop a positive mental attitude*. PositivePsychology.com. https://positivepsychology.com/positive-mindset/#attitude-positive-mindset

Alive Foundation Inc. (2022, May 23). The Importance of Faith. *Morning Coach*. September 28, 2024, https://www.morningcoach.com/blog/the-importance-of-faith#:~:text=Faith%20erases%20doubt%20and%20the,giving%20purpose%20to%20one's%20life.

Baker, R.E., Mahmud, A.S., Miller, I.F. *et al.* Infectious disease in an era of global change. *Nat Rev Microbiol* 20, 193–205 (2022). https://doi.org/10.1038/s41579-021-00639-z

Bio.org. (2022). (rep.). *Measuring Diversity in the Biotech Industry: Tracking Progress in Small and Large Companies*. Bio.org. Retrieved September 17, 2024, from https://www.bio.org/sites/default/files/2022-06/261734_BIO_22_DEI_Report_P4.pdf.

Birt, J. (2023, July 31). *24 reasons why mentorship is important for Mentee and mentor | indeed.com*. Indeed.com .

https://www.indeed.com/career-advice/career-development/why-is-a-mentor-important

College Consensus. (2024, June 20). *10 best degrees for biotech - what to study for a career in biotechnology*. College Consensus. https://www.collegeconsensus.com/degrees/biotech/

Dennison , K. (2023, April 13). *The Importance of Upskilling and Continuous Learning in 2023. Forbes*, Forbes Magazine. https://www.forbes.com/sites/karadennison/2023/04/13/the-importance-of-upskilling-and-continuous-learning-in-2023/.

Harutyunyan, A. (2024, March 26). *Ventures - Biopharma Therapeutics & Platforms – 2023*. DealForma. https://dealforma.com/ventures-biopharma-therapeutics-platforms-2023/

Herrity, J. (2024, August 9). *How to write An action plan (with template and example) | indeed.com*. Indeed.com. https://www.indeed.com/career-advice/career-development/how-to-write-an-action-plan

Hill, N. (2011, May 12). *Amazon.com: Napoleon Hill's Keys to Success: The 17 principles of personal achievement (Audible Audio Edition): Napoleon Hill, Joe Slattery, audible studios: Audible books & originals*. Amazon.com.

https://www.amazon.com/Napoleon-Hills-Keys-to-Success-audiobook/dp/B0050FLPHC

Japan.Gov. (2024, February 29). *Ikigai: The Japanese secret to a joyful life*. The Government of Japan - JapanGov - https://www.japan.go.jp/kizuna/2022/03/ikigai_japanese_secret_to_a_joyful_life.html

Joubert, S. (2024, August 9). *Top biotechnology salaries in 2024*. Graduate Blog. https://graduate.northeastern.edu/resources/biotechnology-salary/

Kayadelen, D. (2020, January 3). *10 ways about how can you break your limits, overcome your fears and go out of your comfort zone?*. Out of Comfort Zone. https://www.outofourcomfortzone.com/post/10-ways-about-how-can-you-break-your-limits-overcome-your-fears-and-go-out-of-your-comfort-zone

Mayo Foundation for Medical Education and Research. (2022, January 12). *The health benefits of good friends*. Mayo Clinic. https://www.mayoclinic.org/healthy-lifestyle/adult-health/in-depth/friendships/art-20044860

Precedence Research. (2024). *Biotechnology market size: Share and trends 2024 to 2034.*

https://www.precedenceresearch.com/biotechnology-market#:~:text=The%20global%20biotechnology%20market%20size%20accounted%20for%20USD%201.55%20trillion,USD%20521.02%20billion%20in%202023.

Sklar, A. (2024, September 20). *7 ways to build a strong professional network.* Business Class: Trends and Insights | American Express. https://www.americanexpress.com/en-us/business/trends-and-insights/articles/7-ways-to-build-a-strong-network/

Suttie, J. (2021, July 5). *What daydreaming does to your mind.* Greater Good . https://greatergood.berkeley.edu/article/item/what_daydreaming_does_to_your_mind

Made in the USA
Coppell, TX
16 March 2026

73997735R00049